SPACE MYSTERIES
WHAT HAPPENS TO SPACE PROBES?

Gareth Stevens
PUBLISHING

BY TODD SWATLING

Please visit our website, www.garethstevens.com. For a free color catalog of all our high-quality books, call toll free 1-800-542-2595 or fax 1-877-542-2596.

Library of Congress Cataloging-in-Publication Data

Names: Swatling, Todd, author.
Title: What happens to space probes? / Todd Swatling.
Description: New York, New York : Gareth Stevens Publishing, [2019] | Series: Space mysteries | Includes glossary and index.
Identifiers: LCCN 2017054373| ISBN 9781538219591 (library bound) | ISBN 9781538219614 (pbk.) | ISBN 9781538219621 (6 pack)
Subjects: LCSH: Space probes--Juvenile literature. | Interplanetary voyages--Juvenile literature. | Planets--Exploration--Juvenile literature. | Outer space--Exploration--Juvenile literature.
Classification: LCC TL795.3 .S93 2018 | DDC 629.43/54--dc23
LC record available at https://lccn.loc.gov/2017054373

First Edition

Published in 2019 by
Gareth Stevens Publishing
111 East 14th Street, Suite 349
New York, NY 10003

Copyright © 2019 Gareth Stevens Publishing

Designer: Katelyn E. Reynolds
Editor: Joan Stoltman

Photo credits: Cover, p. 1 NASA/JPL/UMD; cover, pp. 1, 3-32 (background texture) David M. Schrader/Shutterstock.com; pp. 3-32 (fun fact graphic) © iStockphoto.com/spxChrome; p. 5 (inset) Lauren Elisabeth/Shutterstock.com; pp. 5 (main), 21, 27 (Mars) NASA/JPL-Caltech; p. 7 NASA - NSSDCA; pp. 9, 11, 17, 18, 19 NASA/JPL; p. 13 ESA/NASA; p. 15 (trajectory) Wronkiew/Wikipedia.org; p. 15 (planets) Christos Georghiou/Shutterstock.com; p. 16 NASA; p. 23 NASA/JPL-Caltech/Cornell/Arizona State Univ.; p. 25 USAF 388th Range Sqd/NASA; p. 27 (diagram) Xession/Wikipedia.org; p. 29 Bill Stafford-NASA-JSC.

All rights reserved. No part of this book may be reproduced in any form without permission in writing from the publisher, except by a reviewer.

Printed in the United States of America

CPSIA compliance information: Batch #CS18GS: For further information contact Gareth Stevens, New York, New York at 1-800-542-2595.

CONTENTS

What Are Space Probes? 4

Starting the Space Age 6

Landers 8

Orbiters 10

All Good Orbits Come to an End 12

Flybys 14

The Voyager Probes 16

The Golden Records 18

Phone Home 20

Spirit and Opportunity 22

Smashed Samples 24

Metric Mix-Up 26

A Hard Life for Probes 28

Glossary 30

For More Information 31

Index 32

Words in the glossary appear in bold type the first time they are used in the text.

WHAT ARE SPACE PROBES?

How do we know anything about other **planets** and outer space? Space probes! Space probes are unmanned spacecraft, which means they don't have any people on board to fly them. They have cameras and other tools that collect **data** and **samples** for scientists on Earth.

Probes explore, or search to find out new things, in places people can't go. Scientists have used probes to explore all sorts of places, from the upper parts of Earth's **atmosphere** to the edge of our **solar system**!

You've seen a space probe if you've seen the movie *WALL-E*. EVE was a probe! Space probes in real life aren't quite like that, but they're still cool science machines!

STARTING THE SPACE AGE

The first space probe, Sputnik 1, **launched** from Earth on October 4, 1957. The space age had begun! Sputnik 1 was the first man-made object to **orbit** Earth. It didn't have any tools on board for collecting data, but it proved that messages could be sent from space back to Earth!

Sputnik 1 was in space for 3 months and made more than 1,400 trips around Earth. It traveled about 5 miles (8 km) per second, so each trip only took about 98 minutes!

OUT OF THIS WORLD!

Sputnik 1 was an orbiter! Its **radio signal**—a beeping sound—could be heard on special home radios on Earth as it flew through space!

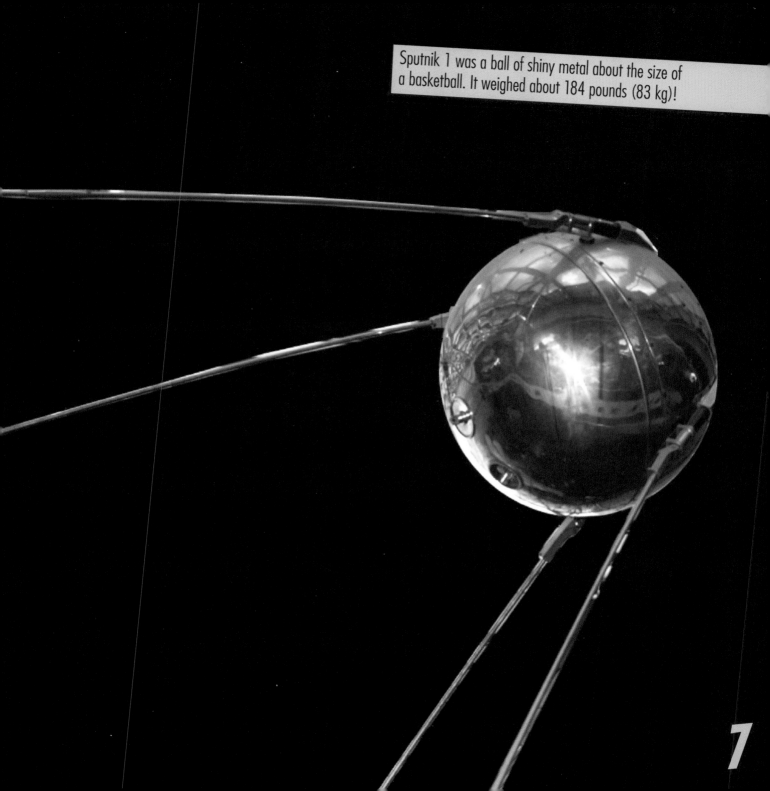

Sputnik 1 was a ball of shiny metal about the size of a basketball. It weighed about 184 pounds (83 kg)!

LANDERS

Probes that can land on objects in space are called landers. Landers use rockets and **parachutes** to slow down to a safe speed before landing.

Some landers have robots called rovers that can drive around on the planet they land on! Rovers have all sorts of science tools for collecting data. Some have arms that can pick up rocks. There's even a special rover tool that can break up rocks to see what they're made of!

OUT OF THIS WORLD!

Landers usually stay in the spot where they land. They'll keep taking pictures and measurements and sending data back to Earth until they run out of power!

This is a picture taken from the Pathfinder lander after it landed on Mars's surface!

Sojourner

Pathfinder

ORBITERS

An orbiter is a probe that orbits an object in space while taking pictures and collecting data. Orbiters can stay in space a long time—the Mars Odyssey probe has been orbiting Mars since October 2001!

Landers and rovers can send messages all the way to Earth, and orbiters often help! Orbiters have longer antennae and more power than the rovers and landers. This means they can send more data back to Earth at faster rates.

This is an artist's idea of what the Odyssey probe looks like while orbiting Mars.

ALL GOOD ORBITS COME TO AN END

An orbiter will lose speed as it bumps into small bits of dust or even air if the object it's orbiting has an atmosphere. When an orbiter slows down, **gravity** pulls it closer to the object it's orbiting until it crashes. This is called an "orbit decay."

When a probe hits air, the air gets compressed, or pressed together. Compressed air quickly heats up everything around it, which can cause the probe to burn up in the atmosphere.

This is what it looks like when something burns up in Earth's atmosphere.

13

FLYBYS

When a probe flies close to the object it's studying without entering orbit or landing on it, it's called a flyby. Probes learn a lot during flybys, and they're great opportunities for pictures!

Probes also do flybys to change speed or direction. Scientists will fly a probe near a planet to use the planet's gravity to change the probe's course. This is called a "gravity assist" flyby. The Rosetta probe used Mars for a gravity assist flyby in 2007.

OUT OF THIS WORLD!
When Voyager 1 did a flyby of Saturn in 1981, it came within 39,892 miles (64,200 km) of the planet's clouds!

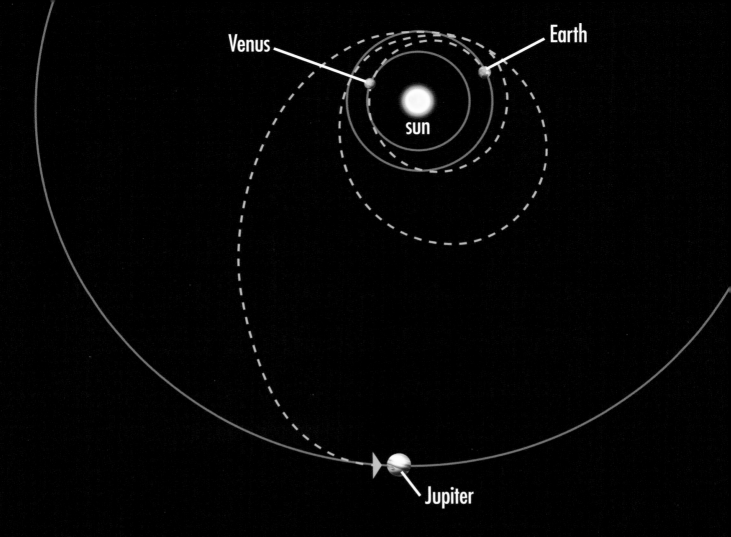

Some probes do a lot of flybys, either to study an object again or to pick up speed. Here's the path the Galileo probe took to Jupiter!

THE VOYAGER PROBES

In 1977, scientists sent two probes—Voyager 1 and 2—to study the farthest parts of the solar system. As they traveled, the probes did flybys of many of the planets in our solar system, taking the first closeup pictures of them!

Forty years and more than 10.8 billion miles (17.4 billion km) later, the Voyager probes are still working! Scientists think they'll run out of power around 2025, but their **mission** won't end there.

Voyager 2 near Saturn

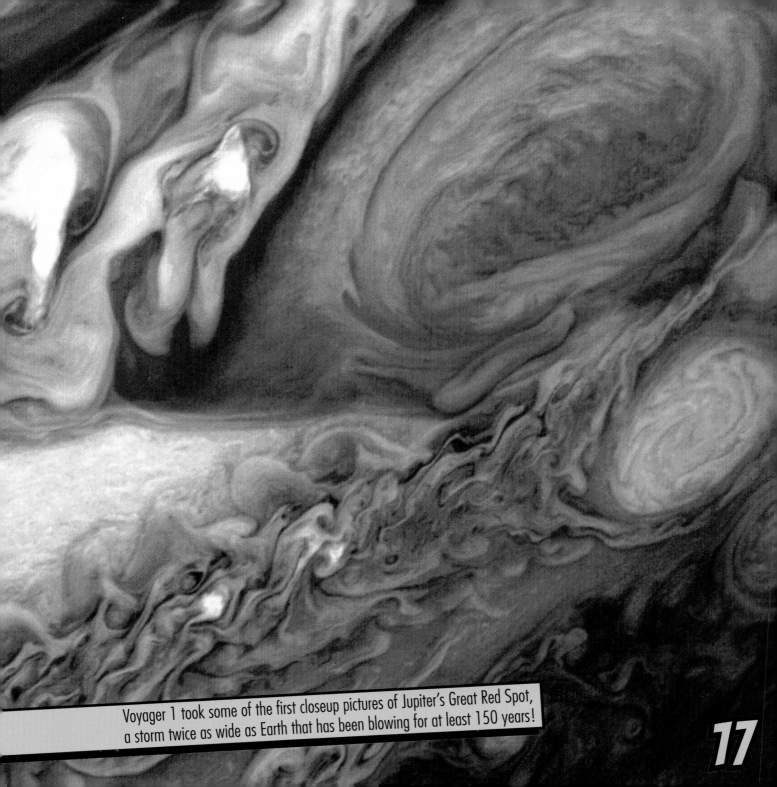

Voyager 1 took some of the first closeup pictures of Jupiter's Great Red Spot, a storm twice as wide as Earth that has been blowing for at least 150 years!

THE GOLDEN RECORDS

Even without power, the Voyager probes will keep shooting through space at about 10 miles (16 km) per second. Space is very empty, so there's little to slow them down!

Scientists knew these two space probes could keep traveling for billions of years! So, they sent messages for whoever—or whatever—might one day discover the spacecraft. The Voyager probes each carry a golden record, or disc that stores data. The records contain sounds and pictures of life on Earth.

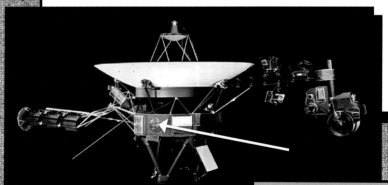

golden record on Voyager probe

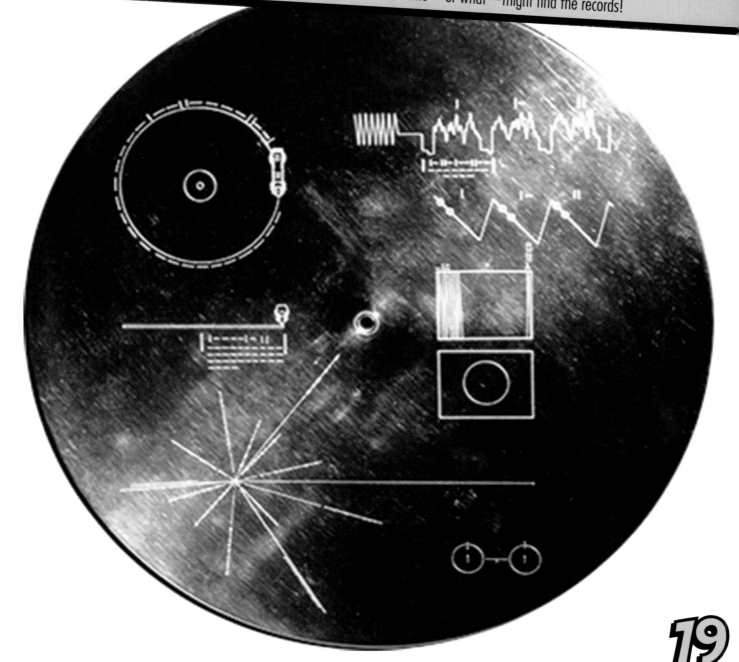

The covers of the golden records have pictures showing how to use them and where they came from. Scientists chose pictures rather than words because they weren't sure who—or what—might find the records!

PHONE HOME

The radio signals through which space probes and scientists on Earth "talk" to each other move at the speed of light. But even at that speed, it can take a long time for messages to be sent and received.

Sometimes space probes can't wait for a message from Earth. A rover might crash into something if it had to wait for a scientist to tell it to stop! So, sometimes space probes "think on their own" for some tasks.

OUT OF THIS WORLD!

Voyager 1 is so far away it takes over a day and a half for a signal to get to Earth and back!

On Mars, the Curiosity rover can pick out interesting rocks on its own. It can even see what they're made of using a laser, which is a tool that uses a special beam of light!

SPIRIT AND OPPORTUNITY

The Spirit and Opportunity rovers landed on Mars in 2004. They were only built to last for 3 months and travel about 0.75 miles (1 km).

Spirit kept rolling for 5 years! It got stuck in sand, but was able to keep collecting data for about another year. Scientists think its solar panels didn't get enough sunlight and it wasn't able to stay warm enough to keep working. Opportunity is still working today, nearly 14 years longer than planned!

OUT OF THIS WORLD!

Solar panels allow probes to make electricity from the sun's light. They use this electricity to power their tools and help them stay warm enough to operate. The average temperature on Mars is around -83°F (-64°C)!

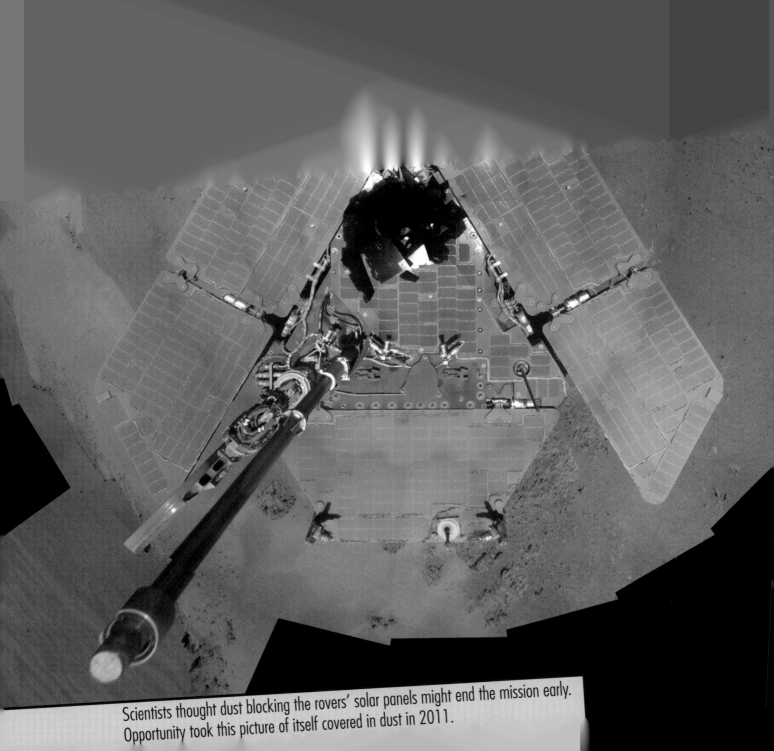

Scientists thought dust blocking the rovers' solar panels might end the mission early. Opportunity took this picture of itself covered in dust in 2011.

SMASHED SAMPLES

In 2001, scientists launched the Genesis probe. Its mission was to gather samples of the tiny **particles** given off by the sun. A mission to bring a space sample to Earth is called a "sample return mission."

Scientists thought a landing slowed by just a parachute might break the sample. They planned to have a helicopter gently catch it after the parachute slowed it down, but the parachute never opened! In 2004, the sample smashed into the ground at 193 miles (311 km) per hour!

OUT OF THIS WORLD!
The Genesis probe's sample weighed only 0.00001411 ounces (0.4 mg)—half the weight of an ant!

A helicopter practiced catching the parachute with a model of Genesis. It worked during practice, but not when it really mattered!

METRIC MIX-UP

In 1999, a probe called the Mars Climate Observer was lost while trying to enter orbit around Mars. The data from the probe looked the way it was supposed to. However, after closer review, scientists learned the probe was much closer to the planet than it should have been.

Scientists soon discovered why the mission failed. While building the probe, one team used metric units for their math, but another team used English units! This meant the probe was about 56 miles (90 km) off course.

OUT OF THIS WORLD!

Nobody knows exactly what happened to the Mars Climate Observer, but it probably burned up in Mars's atmosphere. This was a $125 million mistake!

WAY OFF!

planned path

↕ 56 miles (90 km) apart

actual path

The distance the Mars Climate Observer was off course by is about the same distance between Earth's surface and outer space!

A HARD LIFE FOR PROBES

Space is a hard place to survive. It can quickly become very hot or very cold by hundreds of degrees. There are clouds of dust and rocks speeding through space. Probes naturally start wearing out in these conditions.

Whether they crash, break down, burn up, or just run out of power, probes can't keep working forever. Since we can't yet travel far enough into outer space to bring them **fuel** or fix them, we'll just have to keep building more probes!

OUT OF THIS WORLD!

It takes many people to plan and build a probe and send it into space. Who knows, maybe someday you'll build a space probe!

People can't go to many places in space yet, but probes are helping us learn how one day we might be able to!

GLOSSARY

atmosphere: the mixture of gases that surround a planet

data: facts and figures

fuel: matter that is burned to produce heat or power

gravity: the force that pulls objects toward the center of a planet or star

launch: to send or shoot something into the air or water or into outer space

mission: a task or job a group must perform

orbit: to travel in a circle or oval around something, or the path used to make that trip

parachute: a specially shaped piece of cloth that collects air to slow something down

particle: a very small piece of something

planet: a large, round object in space that travels around a star

radio signal: a message, sound, or picture that is carried by waves of light or sound without using wires

sample: a small part of something

solar system: the sun and all the space objects that orbit it, including the planets and their moons

FOR MORE INFORMATION

BOOKS

Carney, Elizabeth. *Mars*. Washington, DC: National Geographic, 2014.

Kortenkamp, Steve. *Future Explorers: Robots in Space*. North Mankato, MN: Capstone Press, 2016.

Parker, Steve. *The Story of Space: Probes to the Planets*. Mankato, MN: A+ Smart Apple Media, 2016.

WEBSITES

NASA Explore Mars Trek
www.mars.nasa.gov/maps/
See a map of Mars made from pictures taken by Mars probes.

Voyager: Mission Status
voyager.jpl.nasa.gov/where
See exactly how far each Voyager probe is from Earth and which tools on the probes are still working!

Voyager: The Golden Record
www.voyager.jpl.nasa.gov/golden-record/
See and hear what NASA put on Voyager's golden record!

Publisher's note to educators and parents: Our editors have carefully reviewed these websites to ensure that they are suitable for students. Many websites change frequently, however, and we cannot guarantee that a site's future contents will continue to meet our high standards of quality and educational value. Be advised that students should be closely supervised whenever they access the internet.

INDEX

atmosphere 4, 12, 13, 26

Curiosity 21

data 4, 6, 8, 10, 18, 22, 26

Earth 4, 6, 8, 10, 13, 15, 17, 18, 20, 22, 24, 27

flyby 14, 15, 16

Galileo 15

Genesis 24, 25

golden records 18, 19

gravity 12, 14

Jupiter 15, 17

lander 8, 9, 10

Mars 9, 10, 11, 14, 21, 22, 26, 27

Mars Climate Observer 26, 27

Odyssey 10, 11

Opportunity 22, 23

orbit 6, 10, 11, 12, 14, 15, 26

orbiter 6, 10, 12

Pathfinder 9

planets 4, 8, 14, 15, 16, 26

radio signals 6, 20

Rosetta 14

rovers 8, 10, 20, 21, 22, 23

Saturn 14, 16

Sojourner 9

solar panels 22, 23

Spirit 22

Sputnik 1 6, 7

sun 15, 22, 24

Venus 15

Voyager probes 14, 16, 17, 18, 20